Bijoy Boban

An analysis on the dendral expert system

Der GRIN Verlag publiziert seit 1998 wissenschaftliche Arbeiten von Studenten, Hochschullehrern und anderen Akademikern als eBook und gedrucktes Buch. Die Verlagswebsite www.grin.com ist die ideale Plattform zur Veröffentlichung von Hausarbeiten, Abschlussarbeiten, wissenschaftlichen Aufsätzen, Dissertationen und Fachbüchern.

Document Nr. V213082

Bijoy Boban

An analysis on the dendral expert system

GRIN Verlag

Die Deutsche Bibliothek verzeichnet diese Publikation in der Deutschen Nationalbibliografie;
detaillierte bibliografische Daten sind im Internet über http://dnb.d-nb.de/ abrufbar.

1. Auflage 2013
Copyright © 2013 GRIN Verlag GmbH
http://www.grin.com
Druck und Bindung: Books on Demand GmbH, Norderstedt Germany
ISBN 978-3-656-41436-0

Term Paper of Course code: CSE 508, KNOWLEDGE BASED EXPERT SYSTEMS.
Submitted to Asst Prof. Gour Sundar Mitra Thakur

AN ANALYSIS ON THE DENDRAL EXPERT SYSTEM

Er. Bijoy Boban (Reg No: 11200402, Roll No: A09)
Lovely Institute of Technology and Sciences (School: K2, Session: 209)
Lovely Professional University
Phagwara, India-144411
March 2013

Abstract -**In this paper, we do an analysis on the influential pioneer project in the application area of Heuristic programming for experimental analysis in empirical science using IUPAC conventions. The primary aim of the DENDRAL project was to help organic chemists in identifying unknown organic molecules from compounds extracted from known origin that had medicinal or utility value. The process was done by analysing their mass spectra and then undergoing comparative study using knowledge of chemistry. It was done at Stanford University by Edward Feigenbaum, Bruce Buchanan, Joshua Lederberg, and Carl Djerassi. It began in 1965 and spans approximately half the history of AI research. The DENDRAL Project was one of the first large-scale programs to embody the strategy of using detailed, task-specific knowledge about a problem domain as a source of heuristics, and to seek generality through automating the acquisition of such knowledge [1].**

Keywords- **DENDRAL Expert System, Heuristic DENDRAL, Meta-DENDRAL, plan-generate-test paradigm,**

I. INTRODUCTION

The DENDRAL Expert system is considered the pioneer in expert systems because it automated the decision-making process and problem-solving behaviour of organic chemists with the help of Heuristics. There were two sub-programs in the software architecture of DENDRAL, Heuristic DENDRAL and Meta-DENDRAL. It was coded in Lisp (programming language), which was considered the language of AI during that time. Many systems were derived from DENDRAL, including MYCIN, MOLGEN, MACSYMA, PROSPECTOR, XCON, and STEAMER. The name DENDRAL is a portmanteau of the term "Dendritic Algorithm".

The DENDRAL programs were knowledge-driven, in the sense of today's expert systems, with the knowledge principle--that knowledge is power-first articulated in the context of DENDRAL [2]. It was the first of the Expert Systems to use the concept of a separate knowledge base that could be rewritten or redefined for new purposes while having retained all the same source for interpreting and using that information. DENDRAL was the first rule-based system made for a "real-world" problem. It was intended for chemist, other than its developers, for undergoing their research, as this was developed during the cold war it had a lot of other significance too. This project was handled by more than one discipline and was in use for a long time. It was one of the larger more sustained AI projects carried out, making it more prominent other than its successes. Perhaps most significant is that this research was an extensive empirical exploration of heuristic programming techniques as such it was a validation of the strengths and weaknesses of these techniques and an instantiation of a philosophical concept of automatic discovery procedures whose status had long been in dispute [2]. If we go on to define human knowledge as the information that a human being possesses, then Lindsay and Lederberg agree that expert systems, and in particular, DENDRAL, meet that requirement [3]. DENDRAL was built to add up a database of the known rules (valence requirements) and exceptions in organic chemistry that determine the structures of molecules – the data and knowledge a human chemist might possess and use. Therefore, by definition, it possesses and utilizes this human knowledge. I want to go a step further and agree with Lindsay that expert systems, specifically DENDRAL's Meta-DENDRAL, possess the learning aspect of human behaviour. Others may argue that DENDRAL and other expert systems cannot possibly learn as humans do, and do not even represent human knowledge. Lederberg even himself points out in retrospect that DENDRAL had flawed rules and knowledge, and even knowingly left out specific "rules" or data [3].

If we consider human brain having conscious and un conscious mind, then we consider the short term and long term memories also and we can define human brain as voluntary and involuntary too, the reason to take these three set of inter related names is that if we want to link an expert system and an error free human; then an unconscious mind possessing, involuntarily working human brain in a hypnotic state is similar to an expert system, i.e. in that state the humans won't forget to take any link of memory that relates to the question which is asked similar to the Agenda in the inference engine of an expert system. It goes through or fires all the rules that are related and in the memory which is the long term memory that is activated at this point (if it was short term then he will be remembering his state of psychotic sleep) and collects all the five types of memory i.e. sight, sound, touch, smell, taste if exist. If it's a conscious human then there is a gap between the brain memory and the voluntary thinking interface, at that stage humans won't be fully certain about a decision but then comes the most important part, the ability to take risk and the power to handle a new situation. So even the

knowledge engineers of DENDRAL expert system were trying to harvest this state to have a leap to replicate error free expert human thinking.

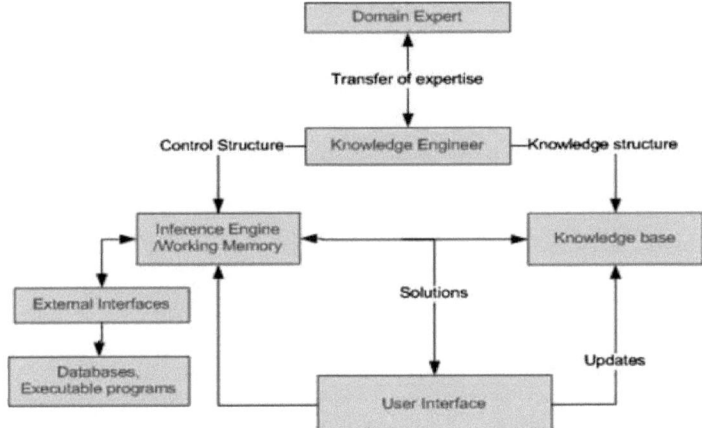

Fig 1: General working of an expert system which can also be seen in DENDRAL.

II. DENDRAL PROJECT ORGANISATION

There have been few successful and long-term interdisciplinary projects in the history of science. We believe DENDRAL should be counted among them. The project worked cohesively for a decade and it involved productive interaction of researchers from the disciplines of chemistry, computer science, genetics, philosophy, physics, mathematics, electrical engineering, management science, and psychology. It is difficult to give a recipe for this success but we believe we can list some important ingredients. The task was conceived in such a way as to appeal to many interests, it could have been described as a pure mass spectrometry problem or a content free hypothesis formation problem but it was not. This task is not prohibitively difficult it can be understood (with a moderate effort) by anyone with a modest technical background. One scientist with knowledge of both chemistry and computer science was willing to coordinate and arbitrate the often conflicting efforts of the group and was able to do it because others felt sufficient respect for his ideas and vision to sacrifice some of the traditional autonomy and rugged individualism of scientists. The project leaders were skilled managers who had learned to delegate responsibility through management of other academic organizations.

They also shared a willingness to take risks with unproven personnel. Not the least important but a natural selection occurred resulting in a staff of specialists each of whom was truly willing to go more than half way to understand the other's discipline, paradigms and arcane jargon. There was also a genuine desire among the computer science personnel to create programs of value to chemists on the way to solving the big problem. Although many of these utilities did not use AI methods they provided tangible benefits to the chemist collaborators whose assistance was essential. This piece of common sense ("quid pro quo") is missing in projects that skim the cream in a new problem area and that have left colleagues disgruntled about AI. We may offer no magic advice here but the lessons are important and mistakes are costly. Interdisciplinary work is antithetical to most scientists no matter how wistfully they long for it. It is expensive folly to establish a project or institute and fill it with scientists from a variety of disciplines selected only on the basis of scientific credentials. Without leadership specific common goals, mutual empathy, human consideration and a great deal of effort the result will be a collection of scientists none of whom has a colleague. Finally, it should be noted that it is not easy to get funds for a large, interdisciplinary project. It is important to find a sponsoring agency that is willing to invest in long-term research, because continuity is critical, the original proposal memo for the DENDRAL expert system is attached in Appendix 1.

III. METHODS

DENDRAL expert system was one of the first systems with which the phrase expert system has been associated. The DENDRAL project commenced in 1965 at Stanford University. The system was developed by J. Lederberg, an organic chemist (and Nobel Prize winner in chemistry), in conjunction with E.A. Feigenbaum and B.G. Buchanan, both well-known research scientists in artificial intelligence at that time. The DENDRAL system was developed to assist in the field of organic chemistry to determine the structural formula of a chemical compound that has been isolated from a given sample. In determining a structural formula, information concerning the chemical formula, such as C_4H_9OH for butanol, and the source the compound has been taken from, is used as well as information that has been obtained by subjecting the compound to physical, chemical and spectrometric tests.

The original DENDRAL algorithm was developed by J. Lederberg for generating all possible isomers of a chemical compound. DENDRAL contains a subsystem, the Structure Generator, which implements the DENDRAL algorithm, but in addition incorporates various heuristic constraints on possible structures, thus reducing the number of alternatives to be considered by the remainder of the system. Heuristic DENDRAL helps in interpreting the patterns in a spectrogram. It contains a considerable amount of chemical knowledge especially organic chemistry. To this end, another subsystem of Heuristic DENDRAL, called the Predictor, suggests expected mass spectrograms for each molecular structure generated by the Structure Generator. Each expected mass spectrogram is then tested against the mass spectrogram observed using some measure of similarity for comparison. This has been implemented in the last part of the system, the Evaluation Function. Usually, more than one molecular structure matches the pattern found in the spectrogram. Therefore, the system usually produces more than one answer, ordered by the amount of evidence favouring them [4].

A heuristic is a rule of thumb, an algorithm that does not guarantee a solution [6], but reduces the number of possible solutions by discarding unlikely, less chance and irrelevant solutions. The use of heuristics to solve problems is called heuristics programming and was used in DENDRAL to allow it to replicate in machines the process through which human experts infers the solution to problems using rules of thumb and specific, proven information. Heuristics programming was a major approach and a giant step forward in artificial intelligence, as it allowed scientists to finally automate certain traits of human intelligence. It became prominent among scientists in the late 1940s through George Polly's book, How to Solve It: A New Aspect of Mathematical Method. As Herbert Simon said in The Sciences of the Artificial, if you take a heuristic conclusion as certain, you may be fooled and disappointed; but if you neglect heuristic conclusions altogether you will make no progress at all.

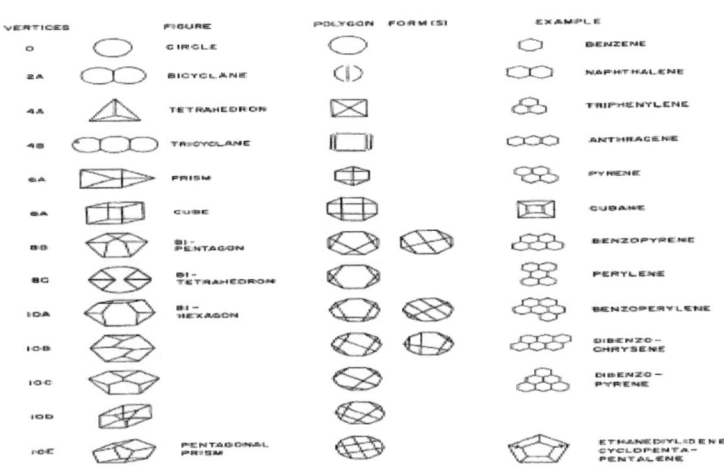

Fig 2: The user interface symbolic output mapping of cyclic chemical structure rendered in DENDRAL.

The primary aim of knowledge engineering is to attain a productive interaction between the available knowledge base and problem solving techniques. This is possible through development of a procedure in which large amounts of task-specific information is encoded into heuristic programs. Thus, the first essential component of knowledge engineering is a large knowledge base. The knowledge base would include specific knowledge about the mass spectrometry technique, large amount of information that forms the basis of chemistry and graph theory, and any information that might be helpful in finding the solution of a particular chemical structure elucidation problem. Through knowledge engineering DENDRAL is able to use this knowledge base to both determine the set of possible chemical structures that correspond to the input data and form new general rules that help it reduce the number of solutions. Thus, at the end the user is now provided with a minimal number of possible solutions, which can now be tested by any non-expert user to find the right solution.

If we try to define human knowledge as explained in the introduction, the information that a human possesses, then Lindsay and Lederberg agree that DENDRAL expert system meet that requirement. DENDRAL was built with a database of the known rules (valence requirements) and exceptions in organic chemistry that helps a human to determine the structures of molecules manually, the data and knowledge a human chemistry domain expert might possess and use. Therefore, by definition, the DENDRAL expert system possesses and utilizes this human knowledge. Let us go a step further and agree with Lindsay that expert systems, specifically DENDRAL's Meta-DENDRAL, possess the learning potential of human involuntary behaviour. Some argue that DENDRAL expert systems cannot possibly learn as humans do, and do not even represent human knowledge due to the extraordinary power of decision making even in an unknown or unattended situation. Lederberg even himself points out in retrospect that DENDRAL had flawed rules and knowledge, and even knowingly left out specific "rules" or data [5].

We can agree that DENDRAL expert system adequately represent human knowledge when there is no adequate information. As stated above, it clearly possesses the knowledge of rules, and the exceptions to those rules, that a domain expert might have. In response to the flawed knowledge argument that Lederberg presents, even humans possess flawed knowledge [5]. Humans sometimes do not have all the data or even intentionally leave out arguable data when trying to solve problems, and similarly, so does DENDRAL, this state can be called as part of human weakness that do exist when a human expert is there as a domain expert and knowledge engineer which to an extend can't be solved.

Now is where the idea of expert systems acting as human learning systems. First of all, just as humans can be made aware of their faulty data and thus change their knowledge accordingly, so can DENDRAL systems. This Expert systems separate the knowledge database part of the system from the problem solving part of the system, and can thus have their "knowledge" or data altered without really changing the system. Thus expert systems can learn new knowledge and apply it in the future. Taking it one step further, Lederberg points out that Meta-DENDRAL can use its knowledge of known mass spectra/structure pairs to infer the structures of unknown mass spectra, just as a human might. This predicted structure can then be tested in the Laboratory [5].

DENDRAL offers the user little assistance with its methods (although some preliminary work on this problem was done). We summarize with some speculations on why DENDRAL is not used more widely:

- ❖ Chemists do not know it exists.
- ❖ The hardware/software are too expensive.
- ❖ The chemist does not wish to invest the time to learn the system.
- ❖ Exhaustive generation is not seen as essential to the structure elucidation problem.
- ❖ An attitude that "That is not the way it is done", or that a tool "not invented here" is not worth using, or that "Machines can't think; that's my job".
- ❖ The chemist and DENDRAL do not collaborate intensively over a long period of time (much of the chemist's time is spent in the lab); so, like a fire alarm, its value is underestimated.
- ❖ DENDRAL is not cost-effective for any single individual (though it may be for an entire company).
- ❖ Chemical and pharmaceutical companies do not cooperate and share resources.
- ❖ Pieces of the DENDRAL system, e.g., structure-matching algorithms, were easier to market than the whole system.
- ❖ The niche that DENDRAL fills is not perceived by chemists as important enough to warrant use of the system [1].

IV. MODULES

The DENDRAL expert system consists of a heuristic DENDRAL and Meta- DENDRAL along with a Plan- Generation- Test paradigm module. Heuristic DENDRAL is a program that uses mass spectra or other experimental data together with knowledge base of chemistry, to produce a set of possible chemical structures that may be responsible for producing the data. A mass spectrum of a compound is produced by a mass

spectrometer, and is used to determine its molecular weight, the sum of the masses of its atomic constituents. For example, the compound water (H_2O) has a molecular weight of 18 since hydrogen has a mass of 1.01 and oxygen 16.00, and its mass spectrum has a peak at 18 units. Heuristic DENDRAL would use this input mass and the knowledge of atomic mass numbers and valence rules, to determine the possible combinations of atomic constituents whose mass would add up to 18. As the weight increases and the molecules become more complex, the number of possible compounds increases drastically. Thus, a program that is able to reduce this number of candidate solutions through the process of hypothesis formation is essential.

Meta- DENDRAL is a knowledge acquisition system that receives the set of possible chemical structures and corresponding mass spectra as input, and proposes a set of hypotheses to explain correlation between some of the proposed structures and the mass spectrum. These hypotheses would be fed back to Heuristic DENDRAL to test their applicability. Thus Heuristic DENDRAL is a performance system and Meta-DENDRAL is a learning system. The program is based on two important features the plan-generate-test paradigm and knowledge engineering.

The plan-generate-test paradigm is the basic organization of the problem solving method and is a common paradigm used by both Heuristic DENDRAL and Meta- DENDRAL systems. The generator generates potential solutions for a particular problem which are then expressed as chemical graphs in DENDRAL. However, this is feasible only when the number of candidate solutions is minimal. When there are large numbers of possible solutions DENDRAL has to find a way to put constraints that rules out large sets of candidate solutions. This is the primary aim of DENDRAL planner, which is a hypothesis-formation program that employs task specific knowledge to find constraints for the generator. Last but not least, the tester analyses each proposed candidate solution and discards those that fail to fulfil certain criteria. This mechanism of plan-generate-test paradigm is what holds DENDRAL together.

V. DENDRAL'S KNOWLEDGE OF CHEMICAL CONCEPTS AND PROCEDURES.

DENDRAL's knowledge base of facts and concepts was complete enough for many tasks, including the substantive one of generating all isomeric structures, but was admittedly incomplete. It was however, intended to be extensible by using rules, lists, tables and values of LISP atoms or attributes. For example only the half-dozen chemical atoms that most frequently occur in organic compounds were known, but any others could easily be added by editing one list of chemical atoms to be considered plus the property list of each new chemical atom. Some of the lists contained names of LISP functions to be used as special purpose constraints (e.g., to see if a proposed structure violated the principle of stability known as Bredt's rule), or as descriptions of processes that occur in mass spectroscopy (e.g., elimination of water). These were more complex procedures than made sense to decompose into productions and that were treated as named primitives by chemists anyway. Perhaps it is obvious but extensibility was considerably easier in the places where we only needed to edit lists and tables. When new LISP functions needed to be created to serve as functional primitives we attempted to generalize the functions so they could be used in other ways and to consolidate old functions with new ones to create reusable code. When completely new capabilities were considered such as generating ringed structures we often found we needed to rethink the design and implementation of large sections of code (Fig 2). This obvious principle evolved from the need to develop a computable theory of stability (BADLIST) interactively and from the difficulties we encountered in changing code to overcome experts criticisms of the programs performance.

(1) Knowledge of chemical graphs:
- ❖ Atom types (C, H, N, 0, P, S), along with essential properties of each atom: valence, atomic weight, isotopes, relative abundance of each isotope.
- ❖ Bond types (single, double, triple, aromatic).
- ❖ *How to:*
  - o Generate all acyclic and/or cyclic isomers (including fused rings, Spiro forms, etc.).
  - o Compute the degree of unsaturation (number of double bonds and rings) from an empirical formula.
  - o Compute the mass of a collection of chemical atoms (empirical formula of a molecular fragment) at high or low resolution, plus masses and relative abundance of collections with isotopic contributions.
  - o Detect topological symmetry in graphs.
  - o Generate all stereo isomers.
  - o Find all cycles in a graph.
  - o Define and name a new sub graph (with specialized editor).

o Restrict generation of structures to those containing [GOODLIST] or not containing [BADLIST] named sub graphs.
o Find all occurrences of arbitrarily complex sub graphs in a graph.
o Find the greatest common sub graph among a set of graphs.
o Label nodes and edges of a graph in all distinct ways, prospectively avoiding symmetric labelling.
o Draw a chemical structure.

(2) Knowledge of chemical stability:
❖ Twenty classes of unstable families defined as acyclic sub graphs [named on BADLIST], others easily defined and added to BADLIST.
❖ Three complex constraints: terpene rule, isoprene rule, Bredt's rule.
❖ *How to:*
o Recognize keto-enol tautomerism (special form of isomerism) and other tautomers when specified.

(3) Knowledge of mass spectrometry:
❖ Distinction between low-resolution and high-resolution spectra.
❖ Distinction between low-voltage and high-voltage spectra.
❖ *How to:*
o Digitize an analogue mass spectrum.
o Find meta-stable peaks in a spectrum.
o Interpret major features of a mass spectrum, using [about a half dozen] rules that are specific fragmentation patterns for a specific family of compounds-rules already defined for alcohols, ethers, thiols, thioethers, amines, ketones, aromatic acids, estrogenic steroids, androstanes, marine sterols [roughly 50 rules in all are defined].
o Predict a "complete" spectrum in which every bond of a molecule breaks and every bond of every resulting fragment breaks (recursively), with isotopic contributions of atoms.
o Constrain the prediction of a "complete" spectrum to plausible breaks, as specified in the half-order theory table [about a dozen named LISP functions in a table], assigning relative measures of significance or likelihood of various processes and assigning placement of charge to one or more resulting fragments from each process.
o Augment the prediction of a mass spectrum with respect to [about 1-31 rules that are general processes common to every family of compounds-rules already defined for hydrocarbon cleavage, Mc Lafferty rearrangements, elimination of water, carbon monoxide, carbon dioxide, or other user-defined neutral species [about a dozen rules defined].

(4) Knowledge of synthetic chemistry:
❖ *How to:*
o Define a new chemical reaction for the program to consider [about a dozen already defined with a specialized editor].
o Reason about plausible biosynthetic pathways from a known starting material in order to restrict a set of candidate structures to those that are plausible products of the starting material [1].

There are many things that DENDRAL does not know, of course. Among missing items are knowledge of three-dimensional geometry beyond stereoisomerism, polymeric structures, quantum chemistry, and many properties of atoms or structures such as electro negativity, dipole moments, molecular susceptibility, melting points, and crystalline forms. In addition, all the knowledge of LISP is presupposed by the DENDRAL programs. For example, arithmetic and set-theoretic operations, symbol manipulation, interpretation of complex procedures, and countless bookkeeping operations. Considerable amounts of code are devoted to keeping track of intermediate results in the overall processing. This specialized bookkeeping knowledge is not very profound, yet it is indispensable for the integration of many complex procedures. Almost all DENDRAL's knowledge is tailored to the task of molecular structure elucidation. Although some pieces such as the representation and graph matcher have found wider applicability, the problem-solving procedures in DENDRAL are still very special-purpose, complex, and voluminous.

VI. CONCLUSION

Although its utility to working chemists has been limited, DENDRAL is well known to computational chemists, who have incorporated many of the pieces of DENDRAL in their own software. As a single software package DENDRAL no longer runs; without an enthusiastic user community no one has invested time enough to maintain it. The major impact of DENDRAL has been on the AI community, where the program is well known and well enough understood to generate some half truths as well as honest lessons. We take pride in knowing that the lessons learned from DENDRAL are now well entrenched in the AI literature and in the design and implementation of expert systems around the world. We hope this paper will correct some of the half-truths. Daniel Dennett suggests that natural (or artificial) minds are viewed by various writers in three different ways [1]. Mind as Crystal is the view that cognitive science should be modelled after physics: the principles of AI would be the wave equation and general relativity equation of the mind. Mind as Chaos is the counsel of despair: the mind, though it may be a mechanism, is so complex and ad hoc, and its course so sensitive to initial conditions that it is essentially without discoverable principles. The third view, the one to which our work subscribes, is Mind as Gadget, which sees AI as engineering, and its products as special-purpose artefacts constructed according to good design principles, with abilities constrained by specialized knowledge acquired through specialized experiences. This is engineering not in the sense of efficiency and cost effectiveness alone, but in the sense of applying good engineering principles to particular tasks-at-hand. It is a view compatible with a world in which minds and brains evolved by an opportunistic, environmentally sensitive process of evolution, rather than springing full blown as the product of either a grand design or a mystical force. DENDRAL clearly is AI in the spirit of engineering gadget in Dennett's non-pejorative sense; most recent AI systems are as well. While DENDRAL cannot play chess, bake a cake, or diagnose septicaemia, it nonetheless embodies general strategies, augmented by specialized knowledge, that give it a measure of intelligence and which can adapt to new information [5]. The general strategies, recombined with other knowledge, have been shown to do well at other tasks. This is the contemporary view of AI, a view that was advanced and illustrated by DENDRAL. It is perhaps the major legacy of this work.

REFERENCES

[1] DENDRAL: a case study of the first expert system for scientific hypothesis formation, by Robert K. Lindsay *University of Michigan, 205 Zina Pitcher Place, Ann Arbor, MI 48109, USA,* Bruce G. Buchanan *Computer Science Department, University of Pittsburgh, Pittsburgh, PA 15260, USA,* Edward A. Feigenbaum *Knowledge Systems Laboratory, Department of Computer Science, Stanford University, Stanford, CA 94305, USA,* Joshua Lederberg *Rockefeller University, New York, NY 10021~6399.*

[2] E.A. Feigenbaum, Artificial intelligence: themes in the second decade, in: A.J.H. Morrell, ed., *information Processing 68, Proceedings of the IFIP Congress* 1968 (North-Holland. Amsterdam, 1968).

[3] www.dod.us/expertsystems/dendralproject

[4] EXPERT SYSTEMS, Peter Lucas, Institute for Computing and Information Sciences, Radboud University, Nijmegen, The Netherlands, E-mail: peterl@cs.ru.nl.

[5] How DENDRAL was conceived and born. Joshua Lederberg, Rockefeller University, New York, N.Y.

[6] Expert Knowledge, Jason Ruchelsman, Stanford University, .035, March 17th, 2004.

[7] .C. Dennett, When philosophers encounter artificial intelligence. *Daedalus* (1988) 283-295.

Appendix 1: The Original Memo submitted for the approval of DENDRAL project.

Appendix:

This memo is an excellent snapshot of the status of AI research
as seen in 1965, and will therefore be appended in its entirety.

STANFORD ARTIFICIAL INTELLIGENCE PROJECT   April 5,1965
Memo No. 30

AN INITIAL PROBLEM STATEMENT FOR A
MACHINE INDUCTION RESEARCH PROJECT

by E. A. Feigenbaum and R. W. Watson

Abstract: A brief description is given of a research project presently getting under way. This project will study induction by machine using organic chemistry as a task area. Topics for graduate student research related to the problem listed.

The research reported here was supported in part by the Advanced Research Projects Agency of the Office of the Secretary of Defense (SD-183).

We are engaged, in conjunction with Professor Lederberg of the medical school, in a research project which offers possibilities for graduate research, both well defined problems suitable for C.S. 239 projects and not so well defined problems suitable for Ph.D thesis topics. In this memorandum we will define the problem briefly and then outline some sugggested projects. If you are interested in any of the projects or topics suggested, or have a topic to suggest related to this project see either of us for further details.

The long range goal of this research is to attempt to come to grips with the problem of induction by machine. That is, how does one build a machine (write a program) which can interact through a suitable interface with its environment and build and improve models of the environment.

The specific task area chosen in which to attack this problem is organic chemistry and in particular, the determination of the structure of organic molecules from mass spectrograph data. The problem presently facing a chemist is roughly the following:

1) A quantity of an organic molecule is supplied to a mass spectrometer.

2) The molecules are bombarded with electrons which break up the molecules into ionized subparts.

3) The mass spectrometer outputs a spectrum (i.e. a distribution of the masses of the subparts).

4) The larggest mass in the distribution which occurs in any quantity above a given noise level is that of the parent molecule.

5) By trying various combinations of atoms the chemist finds molecular compositions which have a mass equal to that determined in 4. If the resolution of the mass spectrometer is fine enough the determination of a unique composition is possible.

6) Once the chemical composition, or possible compositions, of the molecule is determined, the chemist uses various heuristics in conjunction with the mass spectrum to determine the structure of the molecule.

The computer science problem is to automate the above process. At the present time we see the project as progressing in the following stages.

Stage 0 - Display of Chemical Structures

Professor Lederberg has developed a linear notation for organic molecular structures. Further, he has

devised an algorithm which given a chemical composition as an input will produce as an output all topologically unique organic structures corresponding to this composition. The system is called "Dendral" and exists as an Algol program for the B-5000 written by Larry Tessler.

At the present time many of the structures are not chemically meaningful. Therefore, our first task will be to develop a system which will interact with a chemist and the Dendral system and determine rules for chemically meaningful structures. These rules will be automatically incorporated into a "filter" for the Dendral system.

Presently a program for the PDP-1 exists which accepts a linear Dendral string and displays a chemical graph on the Philco scope. The problem then of Stage O is to improve this program and to develop the software for tying it in with Larry Tessler's program through the disc and which will allow us to use LISP on the 7090 from the Philco scopes.

Stage 1 - Chemist at the Philco Keyboard

During Stage 1 we will develop the programming techniques which will allow a dialogue to take place between the chemist and the system for growing the filter on the Dendral output. This system will involve the display of a graph and the chemist's determination of whether or not it is chemically meaningful. The system must then question the chemist to find out what rules the chemist is using for his determinations and accept his answers in a suitable language. In general, the chemist will not be explicitly conscious of the rules he is using and the machine will serve the important function of helping to bring these rules to a precise awareness.

The end result of Stage 1 is that we will have an improved Dendral system and have learned some important and useful computing techniques. An improved Dendral system and associated display should also be of value to those interested in the problems of information retrieval associated with the chemical sciences.

Stage 2 - Mass Spectrograph Analysis

In stage 2 a chemist and a machine interact in real time through the medium of a scope, scope keyboard, typewriter and possibly light pen or tablet. If the machine were used strictly for performing clerical and algorithmic processes the following dialogue would result.

1) The machine would be supplied with the mass spectrum and would display on the scope face a histogram and the chemical composition(s) of the molecule.

2) The chemist using his experience and peripheral information would then input a linear description of a trial structure which would then be displayed on the scope asa chemical graph, or the Dendral system would be invoked to systematically display chemical graphs which correspond to the given composition.

3) The chemist, using his knowledge of likely places for breaks to occur in the above structure when under electron bombardment, would indicate such a break on the graph. The machine would then compute the mass of the subparts and indicate whether or not such a mass exists in the histogram. Or, the chemist would indicate a mass number in the histogram and the machine would indicate whether or not a subgraph exists which has this mass and if it does exist indicate which subgraph it is.

4) The chemist may also want to move various subgraphs from one place to another and then proceed as above. The machine will then compute the linear canonical form of these new graphs and possibly change the display to a canonical form. Further, the Dendral system may be invoked to systematically change a given subgraph.

5) The chemist eventually finds a structure which he hypothesizes as capable of yielding the mass spectrum.

What we want is for the machine to be used not only for clerical work, but more importantly to learn from the chemist's behavior and therefore take over much of the analysis on its own. To this end we visualize the following variation of the above dialogue.

Initially the machine would be input the correct structures corresponding to different chemical compositions. The chemist would then proceed to present an example analysis of this structure in conjunction with its mass spectrum; finally concluding with the known result that the structure could have yielded the given mass spectrum. During this process the machine will probe the chemist for the rules leading to his behavior. The machine will incorporate these rules in a data structure which will allow the machine to perform a similar analysis.

The machine will then be given a chemical structure corresponding to a given mass spectrum and will be asked to proceed on a step by step analysis of its own. The machine will report its "reasoning" to the chemist as it proceeds. When the machine makes an incorrect step the chemist will interrupt and a dialogue will take place until the machine can make the correct step.

Finally, when the machine can correctly analyze structures known to correspond to given mass spectrums the system will be given a composition and the Dendral generator will be invoked to systematically present for analysis possible structures. Then a dialogue of the following type will take place. The machine will proceed with an analysis as far as it is able and then the chemist will take over. As the chemist manipulates the graph with machine aid, the machine queries the chemist for the rules governing his behavior and a dialogue takes place.

Eventually the chemist reaches a hypothesis that the given structure could or could not yield the given mass spectrum. The machine then proceeds to analyze the structure on its own to see if it would reach the same hypothesis. If not, a further dialogue takes place until the machine can reach the hypothesis of the chemist.

When the machine seems adequate at this task we proceed to Stage 3.

Stage 3 - Good Initial Guesses as to Chemical Structure

In stage 2 the man and machine proceeded systematically through the structures produced by Dendral. Clearly for any large structures the number of isomers of a given chemical composition could run into the millions. Therefore, the chemist must make a good initial guess as to a possible structure and only rely on the Dendral generator to modify subgraphs. Again the chemist and system interact, with the machine querying the chemist to determine the rules for proposing initial structures. The procedures to be followed will be similar to those of the previous stage.

Stage 4 - Refinement of the System

When stage 3 is completed the system will be a good mass spectrum analyzer. However, the data structures produced during this stage will be complicated, duplicated and in general unlikely to be optimum. Therefore, the program and associated data structures which result from Stage 3 will be carefully analyzed to determine how to write an efficient compact system and to determine which sections contain general chemical knowledge and which contain knowledge of a specialized character, useful mainly for mass spectrograph analysis. The final efficient program which results will form the software for some experiments to be undertaken by a suggested mars probe and the efficient program minus the specialized structures will form the basis for a system to be applied to some other chemical tasks such as the synthesis of organic molecules.

The following problems suggest themselves as possible research projects.

1. Display Problems:

In order that the display of the chemical graphs be as useful as possible to the chemist, it should

display the graphs in a form as close as possible to that to which the chemist is trained. This task is difficult to do automatically with our present experience. Therefore, one possible approach at this time is to develop a system which automatically displays a graph close to that desired by the chemist and then allows the chemist to manipulate substructures by simple rotations and bond length adjustments. Another possibility is to allow the chemist to "draw" the graph from the keyboard or with a light pen when it is available.

Because of the size limitations of the scope face it will not be possible to display large molecular structures in their entirety. Therefore, it would be useful to have a "window" mechanism which will allow the chemist to study subsections.

Other features are needed which will allow one to save displays, display more than one graph at a time and perform text editing on the linear input. It would also be useful to allow the chemist to build an initial structure and to later make insertions and deletions as well as move a given substructure to another point on the graph.

As the work on the display proceeds feedback from chemists will indicate other useful refinements to the display system.

2. Various programs need to be written which will allow us to use the facilities of the 7090 from the Philco keyboard.

3. Problems relating to Dendral:

Dendral is a system for canonical representation of chemical structures. However, the chemist is usually not trained in this system and would probably find it easier to input a non-canonical linear string. Therefore, it would be of value to have a routine which would convert this string to a canonical one.

Other more abstract problems relating to the Dendral generator are supplied by Professor Lederberg in appendix A.

4. Mass spectrograph analysis problems:

The chemist will want to have a histogram displayed or some display containing equivalent information. The chemist will further want to indicate a given mass number and have the system determine whether or not there is a subgraph with the indicated mass. The work on this problem will lead to abstract on the searching and comparison of list structures.

It will also be of use to the chemist to be able to indicate a given bond as a likely place for a break to have occurred when under electron bombardment and have the system determine if the masses of the subparts are in the distribution. The chemist will also want to be able to invoke the Dendral generator to systematically mark and change subgraphs.

5. The Dendral filter growing problem:

As mentioned before, the Dendral generator will generate all topologically unique structures regardless of whether or not they are chemically meaningful. The problem here is to grow, on-line, a filter which will only allow chemically meaningful structures to be displayed. To solve this problem, techniques need to be developed so that the chemist can be questioned for his rules of chemical meaningfulness and so that his responses can be dynamically incorporated in a changing data structure. Because the chemist will not always give correct rules, methods must be introduced to guard against the possibility of incorrect rules permanently entering the system. Persons interested in natural language and the computer or formal languages may be interested in this phase of the work.

6. Advanced mass spectrograph analysis problems:

Related to the problem above will be the development of techniques which allow the rules supplied by the chemist for analyzing structures to be directly introduced into an internal machine structure. This structure will allow the system to perform the same functions as the chemist and report to the chemist the important stages of its analysis. The detailed problems in this area will only become clear as we proceed.

It would seem to us that the problems related to the display are the most suitable for M.S. projects as they are quite well defined. The more challenging problems related to the Dendral system and filter and Stage 2 would seem to be of the greatest interest to those contemplating doctoral research.

References

Lederberg, J. "Dendral - 64 A System for Computer Construction,
    Enumeration and Notation of Organic Molecules as Free
    Structures and Cyclic Graphs". Interim Report to the
    National Aeronautics and Space Administration, December 15,
    1964. (Available from either author of this memo).

Lederberg, J. "Topological Mapping of Organic Molecules".
    Proc. Nat. Acad. Sci. 53:134-139, 1965. (Available from
    either author of this memo).